my life cycle

Frog

Cherry Lake Press

Published in the United States of America by Cherry Lake Publishing Group
Ann Arbor, Michigan
www.cherrylakepublishing.com

Reading Adviser: Marla Conn, MS, Ed., Literacy specialist, Read-Ability, Inc.
Book Designer: Jennifer Wahi
Illustrator: Jeff Bane

Photo Credits: © Clara Bastian/Shutterstock.com, 5; © Astrid Gast/Shutterstock.com, 7; © Suway wongkham/Shutterstock.com, 9; © Matej Ziak/Shutterstock.com, 11; © Steve Byland/Shutterstock.com, 13; © SIMON SHIM/Shutterstock.com, 15; © sasapanchenko/iStock.com, 17; © Michal Dobes com/Shutterstock.com, 19; © IRINA ORLOVA/Shutterstock.com, 21; © Joanna K-V/Shutterstock.com, 23; Cover, 2-3, 8, 16, 22, 24, Jeff Bane

Cherry Lake Press is an imprint of Cherry Lake Publishing Group.

Library of Congress Cataloging-in-Publication Data

Names: Gray, Susan Heinrichs, author. | Bane, Jeff, 1957- illustrator.
Title: Frog / Susan H. Gray ; illustrated by Jeff Bane.
Description: Ann Arbor, Michigan : Cherry Lake Publishing, 2021. | Series: My life cycle | Includes index. | Audience: Grades 2-3
Identifiers: LCCN 2020030611 (print) | LCCN 2020030612 (ebook) | ISBN 9781534180055 (hardcover) | ISBN 9781534181762 (paperback) | ISBN 9781534181069 (pdf) | ISBN 9781534182776 (ebook)
Subjects: LCSH: Frogs--Life cycles--Juvenile literature.
Classification: LCC QL668.E2 G675 2021 (print) | LCC QL668.E2 (ebook) | DDC 597.8/9156--dc23
LC record available at https://lccn.loc.gov/2020030611
LC ebook record available at https://lccn.loc.gov/2020030612

Printed in the United States of America
Corporate Graphics

table of contents

About the author: Susan H. Gray has a master's degree in zoology. She loves writing science books, especially about animals. Susan lives in Arkansas with her husband, Michael. Every spring, they hear frogs singing in their neighborhood.

About the illustrator: Jeff Bane and his two business partners own a studio along the American River in Folsom, California, home of the 1849 Gold Rush. When Jeff's not sketching or illustrating for clients, he's either swimming or kayaking in the river to relax.

My life begins in an egg. My mom laid hundreds of eggs in a pond. We stick together in one big **clump**.

In the egg, I **develop** into a **tadpole**. I have a round body and a long tail. I spin and flip around inside my egg. One day, I just pop out!

Soon, I'm swimming around and eating pond plants. I can breathe underwater, like a fish.

One day, my back legs begin growing out. Next, my front legs **sprout**.

Everything is changing! My mouth grows wide, and my eyes **bulge**. My tail shrinks too.

I’ve become a **froglet**. I can breathe the air. My tail is disappearing.

I'm eating bugs now and getting really big. I am becoming an adult.

One day, I hear a male frog calling out for a **mate**.

I find him. He stays with me while I lay eggs.

Those eggs are all in a clump. But soon, new tadpoles will be everywhere!

glossary & index

glossary

bulge (BUHLJ) to grow outward

clump (KLUHMP) a small group of things stuck together

develop (dih-VEL-uhp) to grow and become more complex

froglet (FRAWG-luht) a baby frog that still has a tail

mate (MATE) a partner to produce babies with

sprout (SPROUT) to begin to grow

tadpole (TAD-pole) a very young form of a frog or toad that breathes underwater

index